FREAKY TRUE SCIENCE

MORE FREAKY SPACE STORIES

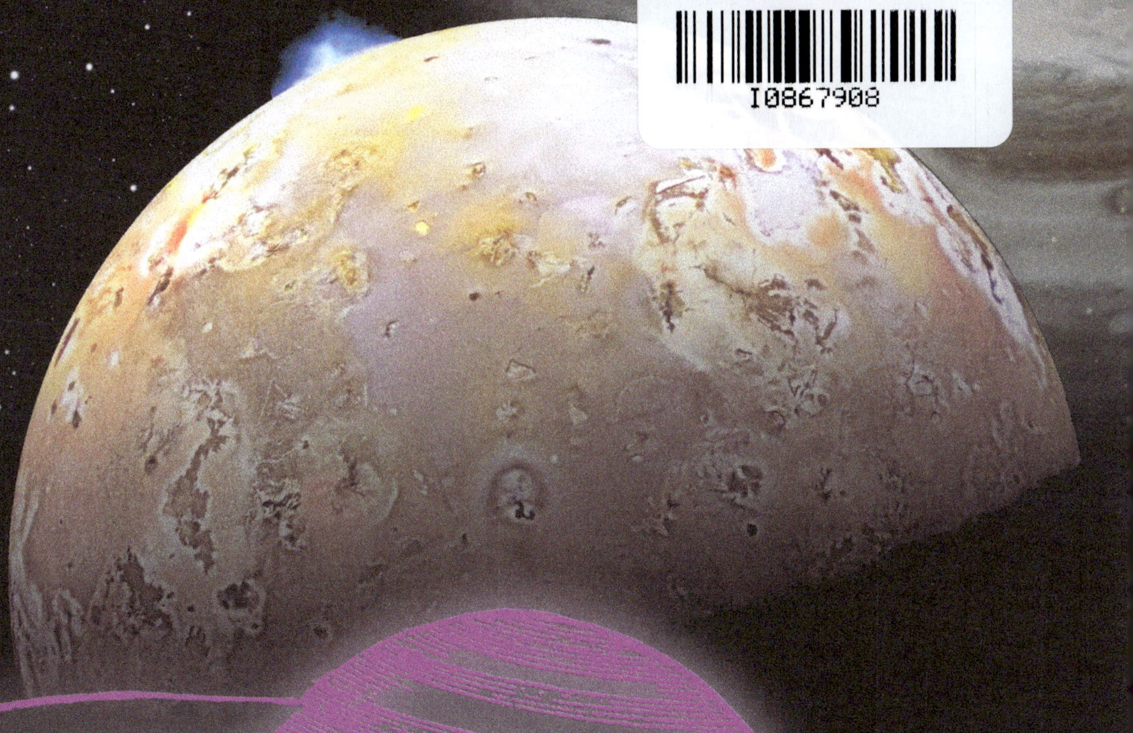

BY ERIC KEPPELER

Gareth Stevens PUBLISHING

Please visit our website, www.garethstevens.com. For a free color catalog of all our high-quality books, call toll free 1-800-542-2595 or fax 1-877-542-2596.

Library of Congress Cataloging-in-Publication Data

Names: Keppeler, Eric.
Title: More freaky space stories / Eric Keppeler.
Description: New York : Gareth Stevens Publishing, 2020. | Series: Freaky true science | Includes glossary and index.
Identifiers: ISBN 9781538240625 (pbk.) | ISBN 9781538240649 (library bound) | ISBN 9781538240632 (6 pack)
Subjects: LCSH: Astronomy–Miscellanea–Juvenile literature. | Space sciences–Miscellanea–Juvenile literature.
Classification: LCC QB46.K477 2020 | DDC 520–dc23

First Edition

Published in 2020 by
Gareth Stevens Publishing
111 East 14th Street, Suite 349
New York, NY 10003

Designer: Sarah Liddell
Editor: Therese Shea

Photo credits: Cover, p. 1 (Io), p. 23 (Kepler Space Telescope) Uwe W./Wikimedia Commons; cover, p. 1 (Jupiter) MarcelClemens/Shutterstock.com; cover, p. 1 (star background) Jurik Peter/Shutterstock.com; cover, p. 1 (moon and Saturn used throughout book) Morphart Creation/Shutterstock.com; background used throughout Shay Yacobinski/Shutterstock.com; hand used throughout Helena Ohman/Shutterstock.com; paper texture used throughout Alex Gontar/Shutterstock.com; p. 5 photo courtesy of ESA/Hubble & NASA; p. 7 NASA/Handout/Hulton Archive/Getty Images; p. 9 muratart/Shutterstock.com; p. 11 oorka/Shutterstock.com; p. 13 NASA images/Shutterstock.com; p. 15 KEES VEENENBOS/SCIENCE PHOTO LIBRARY/Science Photo Library/Getty Images; p. 17 MPI/Stringer/Archive Photos/Getty Images; p. 19 NASA/SCIENCE PHOTO LIBRARY/Science Photo Library/Getty Images; p. 21 Vadim Sadovski/Shutterstock.com; p. 23 (planets) photo courtesy of NASA/Ames/JPL-Caltech; p. 25 James Balog/The Image Bank/Getty Images; p. 27 photo courtesy of Tod Strohmayer (GSFC), CXC, NASA; p. 29 Handout/Handout/Getty Images News/Getty Images.

Printed in the United States of America

CPSIA compliance information: Batch #CS19GS: For further information contact Gareth Stevens, New York, New York at 1-800-542-2595.

CONTENTS

Words in the glossary appear in **bold** type the first time they are used in the text.

WHAT'S OUT THERE?

People have always been curious about outer space, ever since they noticed the beautiful lights flickering in the night sky. At first, they didn't understand the idea of space at all. For example, many once believed that the stars were actually just light from heaven shining through holes in the sky!

But since people have started to learn about space, our understanding has grown immensely. However, so much is still unknown. Outer space is so huge and so full of strange planets, stars, and other objects that we've realized we're only just starting to comprehend it. And the more we learn about what's out there, the more we realize that some of those things are both fascinating and truly freaky!

FREAKY FACTS!

The new eye in the sky from NASA (National Aeronautics and Space Administration) will be the James Webb Space Telescope. Scheduled to launch in 2021, it will be able to view 100 objects at the same time.

THE ANDROMEDA GALAXY IS ONE OF THE MILKY WAY'S NEIGHBORS. IT'S 2.3 MILLION LIGHT-YEARS AWAY. IT CAN BE SEEN ON SOME NIGHTS WITH JUST THE NAKED EYE.

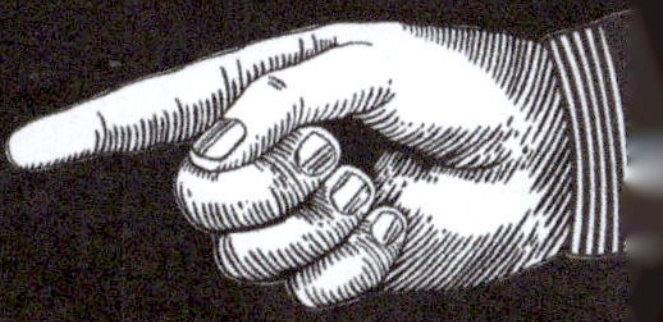

BILLIONS AND BILLIONS

Our solar system is in a **galaxy** we call the Milky Way. The galaxy is so huge that it's very hard for scientists to count all of its stars. The European Space Agency's current estimate is that there are about 100 billion stars in the galaxy, although some believe there could be as many as 700 billion! Scientists also think there are between 100 billion and 200 billion galaxies in the universe.

ALIEN CONTACT?

Astronaut John Glenn was the first American to orbit Earth. He rocketed into space on February 20, 1962, for three trips around the planet. During one orbit, Glenn reported seeing a swarm of tiny glowing particles in space around his ship. Some people thought Glenn imagined them. Others believed the particles were aliens!

"I never saw anything like it. They round a little; they're coming by the **capsule**, and they look like little stars," Glenn reported.

The next astronaut to orbit Earth was Scott Carpenter 3 months later—and he saw the little lights, too! Carpenter figured out that the lights were little pieces of frost on the outside of the spaceship. They appeared to be glowing because they were reflecting sunlight.

FREAKY FACTS!

Glenn's craft was named *Friendship 7*. It went as fast as 17,500 miles (28,164 km) per hour. It covered a distance of 75,679 miles (121,794 km) in less than 5 hours.

WHAT CAUSED THE LIGHTS?

The tiny bits of frost were formed by **condensation** on the outside of the spacecraft. Moisture froze to the craft when it traveled out of the sunlight, but the frost softened and broke loose when the craft came back into the sunlight—causing the glowing particles. In fact, Carpenter was able to make more particles by banging against the inside of his craft, knocking more pieces of frost loose into space.

JOHN GLENN LOOKS OUT THE WINDOW OF HIS SPACECRAFT.

OUR COOL, CLOSE NEIGHBOR

If you look up in the sky at night, probably the easiest thing to spot will be Earth's moon. It orbits our planet and can be seen when sunlight shines on it, day or night.

Scientists think Earth's moon formed about 4.5 billion years ago. Its average distance from Earth is 238,855 miles (384,400 km), and it's only about a quarter as large as our planet. Because it's so much smaller, the moon's gravity is much less than Earth's. So, everything weighs less there than it does here.

FREAKY FACTS!

As it orbits Earth, the moon travels through space at an average velocity of 2,287 miles (3,681 km) per hour!

Here are some freaky moon numbers: Days on the moon are very long—more than 27 Earth days! It gets really hot *and* really cold. Temperatures average 224°F (107°C) during the day and –298°F (–183°C) at night!

SOMETIMES, PEOPLE TELL CHILDREN THE MOON IS MADE OF CHEESE. THIS IS FREAKY—AND UNTRUE. THE MOON IS MADE UP MOSTLY OF ROCK.

SOLAR ECLIPSE

LIGHTS OUT!

Sometimes the moon passes between Earth and the sun, casting its shadow over our planet and briefly making it appear dark—even in the middle of the day! This is called a solar eclipse. Other times, the moon moves directly behind Earth, and our planet's shadow prevents some or all sunlight from reflecting off the surface of the moon. This is called a lunar eclipse. Eclipses used to be scary to witness before astronomers, or people who study the universe, discovered what was happening!

HERE COMES THE SUN!

The sun is the star in the center of our solar system. It provides all the heat and light for the planets around it. And it's huge. How huge? If the sun were as big as the front door on a house, Earth would be about as big as a nickel. Are you ready for another freaky fact? More than 1 million Earths could fit inside the sun!

You couldn't stand on the sun. It isn't solid as our planet is. It's made up almost completely of gases called hydrogen and helium, and it's held together by its own gravity. The outside part of the sun is about 10,000°F (5,500°C), while the **core** is at least 27 million°F (15 million°C)! Scientists think the sun is around 4.5 billion years old.

FREAKY FACTS!

NASA's Parker Solar Probe came within 15 million miles (24 million km) of the sun in November 2018 and will get even closer. The probe reached a record speed of 213,200 miles (343,112 km) per hour.

THIS ILLUSTRATION COMPARES THE SIZE OF THE SUN TO ITS PLANETS. PICTURED FROM LEFT TO RIGHT ARE THE SUN, MERCURY, VENUS, EARTH, MARS, JUPITER, SATURN, URANUS, AND NEPTUNE.

HOW LONG WILL IT BURN?

Scientists think the sun will keep burning for about another 6.5 billion years before it runs out of energy. When it starts to die, it will get so big it will **engulf** Mercury and Venus—and perhaps Earth! The sun will stay that way for as long as 1 billion years and then shrink into what's known as a white dwarf star. When the sun no longer gives off any energy, it will become a black dwarf star.

YOU WOULDN'T WANT TO LIVE HERE

Venus is the second planet from the sun in our solar system. It's 67 million miles (108 million km) from the sun. It's the hottest planet in the whole solar system, even though Mercury is closer to the sun. That's weird, right? This is because Venus has a thick, cloudy atmosphere that traps the heat. Temperatures on Venus can reach 870°F (465°C), which is hot enough to melt the metal lead!

Venus is roughly the same size as Earth. Its surface has mountains and volcanoes. The Maxwell mountain range is 540 miles (870 km) long and is the planet's tallest feature at 7 miles (11 km) high!

Another strange fact about Venus is that it spins the opposite direction than Earth does. So, the sun rises in the west and sets in the east!

FREAKY FACTS!

Scientists think there may once have been water on Venus, though there isn't any now. The planet probably doesn't have life as we know it, but it may have had it long ago!

VISITING VENUS

The first spacecraft to reach Venus was NASA's Mariner 2 on December 14, 1962. Since then, more than 20 spacecraft have explored the planet. Venera 7 from the Soviet Union became the first spacecraft to land on another planet, when it touched down on Venus on December 15, 1970. It lasted just 23 minutes before heat and pressure destroyed it. Japan's Akatsuki probe began orbiting Venus in 2015 and is still working as of early 2019.

THE PLANET VENUS IS NAMED AFTER THE ROMAN GODDESS OF BEAUTY AND LOVE. IT'S VISIBLE WITH THE NAKED EYE AND IS THE BRIGHTEST OBJECT IN THE NIGHT SKY BESIDES THE MOON.

A GIANT SPACE VOLCANO

Mars is the fourth planet from the sun in our solar system. It's the second-smallest planet, but it has the highest mountain in the entire solar system—a volcano called Olympus Mons. This volcano rises 16 miles (26 km) above the ground and stretches over an area roughly 374 miles (602 km) in **diameter**. That's about as big as the state of Arizona!

By comparison, the tallest volcano on Earth is Hawaii's Mauna Loa. Starting at the bottom of the Pacific Ocean, it's 6.3 miles (10 km) high—although only 2.6 miles (4.2 km) above sea level. The highest mountain on Earth above sea level is Mount Everest at 5.5 miles (8.9 km)—still much shorter than Olympus Mons.

FREAKY FACTS!

Mars has two small moons, which were discovered in 1877 by astronomer Asaph Hall. He named them Phobos (which means "fear" in ancient Greek) and Deimos (which means "panic" in ancient Greek).

ASTRONOMERS OBSERVED OLYMPUS MONS FOR HUNDREDS OF YEARS, BUT IT WASN'T IDENTIFIED AS A VOLCANO UNTIL 1971, WITH THE HELP OF THE AMERICAN SPACE PROBE MARINER 9.

A SHIELD VOLCANO

Olympus Mons is a type of volcano called a shield volcano. It's called that because it looks like a large **dome**-shaped shield high above the ground. Scientists think it may have become so large because the gravity on Mars is much less than on Earth. Also, the volcano erupted often, causing **lava** to build up inside. Olympus Mons hasn't erupted in about 25 million years, but it's considered a young volcano. Scientists think it could someday erupt again.

ON EARTH'S LARGEST NEIGHBOR

Jupiter is the fifth planet from the sun in our solar system and a giant compared to other planets. It's two and a half times heavier than all the rest of the planets in our solar system—put together! It's so big that you can sometimes see it with your naked eye, even though it's very far away. Over 1,000 Earths could fit inside of Jupiter!

Jupiter looks much the same all over—except for a large area called the Great Red Spot. It's a gigantic storm with winds as powerful as 400 miles (644 km) per hour! The size and shape of the red spot changes, but it's huge, with a diameter larger than Earth's. Why this freaky storm appears red is still a mystery to scientists!

FREAKY FACTS!

The Great Red Spot has been shrinking over the years. This giant storm may disappear altogether someday!

GREAT RED HOT SPOT

The Great Red Spot is much hotter than most other parts of the colossal planet. Temperatures in Jupiter's atmosphere are around 1,700°F (927°C), while the area above the Great Red Spot is about 2,420°F (1,327°C)—that's hotter than lava! Scientists think the region is so much hotter because the air is heated by both gravity waves and sound waves caused by the violent, stormy conditions of the Great Red Spot.

NO ONE KNOWS HOW LONG THE GREAT RED SPOT ON JUPITER HAS EXISTED. GIOVANNI CASSINI AND ROBERT HOOKE FIRST OBSERVED IT AROUND 1665.

MOON OF FIRE AND ICE

Can you imagine looking up and seeing 79 moons in the night sky? You would if you could stand on Jupiter. Freaky! And Jupiter may have more moons that haven't been discovered yet.

Jupiter's fifth moon is called Io. Io has the most volcanoes of any moon or planet in the whole solar system with around 400—and about 150 are erupting. The volcanoes are located all over the moon and can shoot lava as high as 250 miles (402 km) into space! Temperatures are wildly different on the surface of Io, which is mostly made of rock with seas of fiery lava. While temperatures average –202°F (–130°C), the volcanoes can be as hot as 3,000°F (1,649°C). That's why people call Io a moon of fire and ice.

FREAKY FACTS!

Scientists think Io's volcanic activity is a result of the moon being stretched and squeezed by gravity as it orbits Jupiter.

THE VOLCANOES ON IO WERE FIRST **CONFIRMED** BY NASA'S VOYAGER 1 SPACE PROBE IN 1979. THESE IMAGES OF JUPITER AND ITS MOONS WERE CAPTURED BY THE PROBE.

AN EARLY DISCOVERY

Io is a little smaller than Earth's moon. It's Jupiter's third largest moon and the fifth closest to the planet. Jupiter's four largest moons are Ganymede, Callisto, Io, and Europa. Italian astronomer Galileo Galilei first discovered them on January 8, 1610, though they weren't named until years later. These moons and their movements eventually helped astronomers understand that planets orbit the sun and not Earth.

RINGS IN SPACE

Saturn is the sixth planet from the sun in our solar system. It has at least 62 moons—but its freakiest and most famous features are the bright rings that circle it. Scientists aren't sure how the rings were created, but they're huge. They're about 240,000 miles (386,243 km) wide. That's as far as the distance between Earth and its moon.

It may look like Saturn has just a few rings, but it actually has hundreds of thin rings. If the planet is tilted a certain way, the rings can be hard to see because they're only from 30 to 300 feet (9 to 91 m) thick. They're made up of chunks of rocks and ice, some as small as a grain of sand and some as big as a house!

FREAKY FACTS!

Saturn is so huge that it could fit more than 700 Earths inside!

STILL STUDYING THE RINGS

In 1610, Galileo Galilei became the first to see Saturn through a telescope. He thought the rings were giant moons on either side of the planet, though. Forty-five years later, Dutch scientist Christiaan Huygens figured out what the rings are. Some NASA probes have flown past Saturn, including Pioneer 11 and Voyager 1 and 2. The Cassini spacecraft orbited Saturn for 13 years before crashing into the planet in 2017.

SATURN ISN'T THE ONLY PLANET THAT HAS RINGS! JUPITER, URANUS, AND NEPTUNE ALSO HAVE THEM, BUT THEY'RE MUCH SMALLER AND HARDER TO SEE.

THE GREAT PLANET HUNTER

There are billions of stars and planets in our galaxy, and nothing has been better at finding new ones than the Kepler Space Telescope. Launched in March 2009 at a cost of $700 million, Kepler's mission was to learn if there are other planets like Earth in our galaxy—even ones so far away they can't be seen from the ground.

After nearly 10 years in space, Kepler has finally run out of fuel. NASA had to shut it down in November 2018—but not before it had found 530,506 stars and confirmed the existence of 2,662 planets. What's freaky about that? Some of these planets are like Earth, leading scientists to think there might be alien life on them!

FREAKY FACTS!

The Kepler Space Telescope can't be refueled, but it will continue to orbit the sun for a long time. Right now, it's about 94 million miles (151 million km) from Earth.

BELOW ARE ARTISTS' DRAWINGS OF PLANETS LOCATED BY THE KEPLER SPACE TELESCOPE AND AN ACTUAL PHOTO OF EARTH. THESE NEWLY DISCOVERED PLANETS COULD POSSIBLY SUPPORT LIFE. THEY'RE PICTURED NEXT TO EARTH TO SHOW THEIR SIZE.

ANOTHER EARTH? A LOOK AT KEPLER'S DISCOVERIES

WHO WAS KEPLER?

Johannes Kepler was a German scientist born in 1571. Kepler studied the ideas of early astronomer Nicolaus Copernicus. Kepler came up with three laws of planetary motion. The first law stated planets move in ellipses, or ovals, rather than circles as astronomers of the time thought. Kepler's laws helped persuade people that Earth and the other planets really do move around the sun, as Copernicus had believed.

SEE YOU IN 75 YEARS

People once thought comets were a sign of something terrible or wonderful approaching. A comet may look like a fuzzy star with a freaky tail, but a comet is just a big chunk of rock and ice. When a comet's orbit brings it close to the sun, the sun's heat causes gas and dust to build up around the comet's nucleus, or core. This gassy buildup streams into a tail the nearer the comet gets to the sun.

The most famous comet is Halley's comet. About every 75 or 76 years, it passes our planet and can be seen from the ground. Halley's comet was first observed in 240 BC. The last time it appeared near Earth was in 1986. Scientists expect that it will be back again around 2061.

FREAKY FACTS!

When it passed Earth in 1910, Halley's comet's speed was measured to be 157,838 miles (254,016 km) per hour! This was the first year the comet was caught on film.

WHO WAS HALLEY?

English astronomer Edmond Halley was born in 1656. After studying historical records of comets that appeared near Earth between 1337 and 1698, Halley was the first person to realize that it was the same comet appearing over and over again. He figured out its path in 1705 and correctly guessed that it would appear again in 1758. He didn't get to see the comet again—he died in 1742—but other scientists named the comet in his honor.

THE BODY OF HALLEY'S COMET ISN'T CONSIDERED VERY BIG—IN SPACE TERMS! ITS NUCLEUS MEASURES ABOUT 9 BY 5 MILES (14 BY 8 KM).

A HIDDEN SUN?

Imagine how bright our world could be with two suns! Sometimes, a solar system can have more than one star. A system with two stars is called binary. The two stars orbit a center point. Binary stars can be far apart, but others are so close together that the stars' mass moves back and forth between them!

Some scientists in the 1980s believed that our own solar system might be binary. They thought there might be a star they named Nemesis hiding somewhere in the system. NASA started an exploration program in 2010 to more thoroughly map our solar system. Scientists think this program would have found a second star if it existed—so they mostly agree now that our solar system isn't binary.

FREAKY FACTS!

A binary star system called J0806, which is about 1,600 light-years away, has two stars that orbit each other every 321 seconds!

STAR WARS SCIENCE

In the *Star Wars* movies, Luke Skywalker's home planet of Tatooine is in a system with two stars! Scientists have been able to find some binary stars with one planet—but they didn't find one with more than one planet until 2012. That's when the Kepler Space Telescope discovered two worlds orbiting two stars. Scientists named this star system Kepler-47 after the famous telescope that discovered it.

SOMETIMES BINARY STARS ARE SO CLOSE TOGETHER THAT THEY LOOK FROM EARTH LIKE JUST ONE BIG, BRIGHT STAR.

AND SO MUCH MORE

English astronomer Sir Arthur Eddington once said: "Not only is the universe stranger than we imagine, it is stranger than we can imagine."

Outer space is full of interesting and sometimes strange things. And we only know about the ones we've found so far. But that's just a tiny fraction of what's out there in the universe, so that's why it's important to keep learning more about it.

The more we explore, the more planets and stars we'll find—and maybe some of those other worlds will have people on them! Perhaps they'll be like Earth or maybe they'll be totally different. The only way to find out for sure is to keep exploring the mysteries of space.

FREAKY FACTS!

NASA's New Horizons spacecraft spotted a strange object that looks a bit like a snowman in the outer reaches of our solar system in January 2019. They named it "Ultima Thule," which means "beyond the known world."

ULTIMA THULE, SHOWN HERE, WAS PROBABLY TWO OBJECTS THAT BECAME ONE. IT MAY BE FROM OUR SOLAR SYSTEM'S BEGINNINGS!

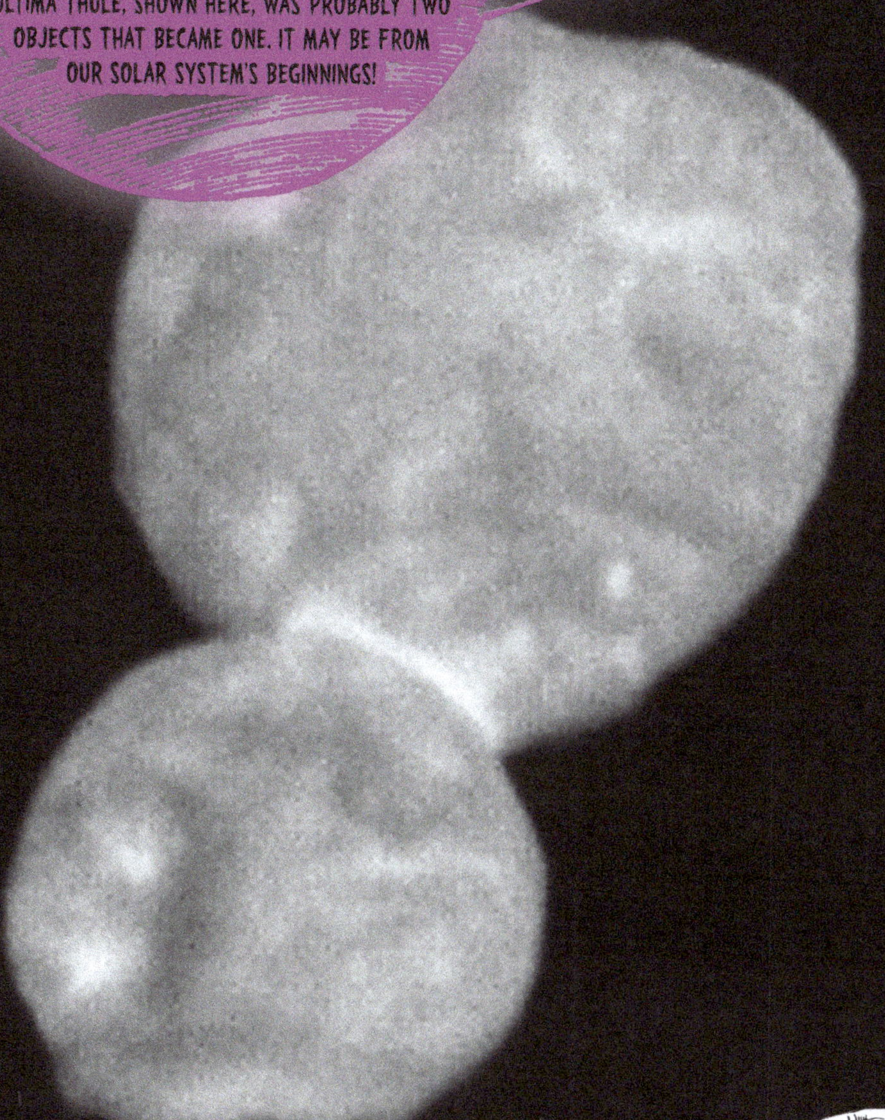

MOVING ON TO MARS

Since Mars is a close planet to Earth, scientists think that could be the next place to send astronauts. But first, they're sending out probes to learn all they can so people will know what to expect when they get there. NASA has successfully landed four robot **rovers** on Mars since 1997 to gather information, with a fifth expected to arrive in 2021. Rovers have wheels and can drive around to explore different parts of the planet.

GLOSSARY

capsule: a small part of a spacecraft that is separate from the rest of the spacecraft and that is where people live and work

condensation: the act or process of a gas or vapor becoming a liquid form. Also, the liquid that forms from this process.

confirm: to show that something is correct

core: the central part of something

diameter: the distance from one side of a round object to another through its center

dome: a large rounded roof or ceiling that is shaped like half of a ball

engulf: to flow over and cover

galaxy: a large group of stars, planets, gas, and dust that form a unit within the universe

lava: melted rock from a volcano

probe: an unmanned spacecraft

rover: a vehicle for exploring the surface of a moon, planet, or other space object

velocity: the speed at which something moves in a particular direction

FOR MORE INFORMATION

BOOKS

Holland, Simon. *Space.* New York, NY: DK Publishing, 2016.

Royston, Angela. *Space.* Mankato, MN: Smart Apple Media, 2016.

Russo, Kristin J. *Space.* North Mankato, MN: Capstone Press, 2018.

WEBSITES

Kids Astronomy
kidsastronomy.com
Find many links to space subjects here.

NASA Space Place
spaceplace.nasa.gov
Learn about outer space from the experts at NASA.

Publisher's note to educators and parents: Our editors have carefully reviewed these websites to ensure that they are suitable for students. Many websites change frequently, however, and we cannot guarantee that a site's future contents will continue to meet our high standards of quality and educational value. Be advised that students should be closely supervised whenever they access the internet.

INDEX